Originally published in 1932

ISBN: 978-1-940001-21-0

Published by Sportsman's Vintage Press
2014

MANUAL

OF

POLICE REVOLVER INSTRUCTION

1st Printing, August, 1932
2d Printing, July, 1933
3d Printing, October, 1934

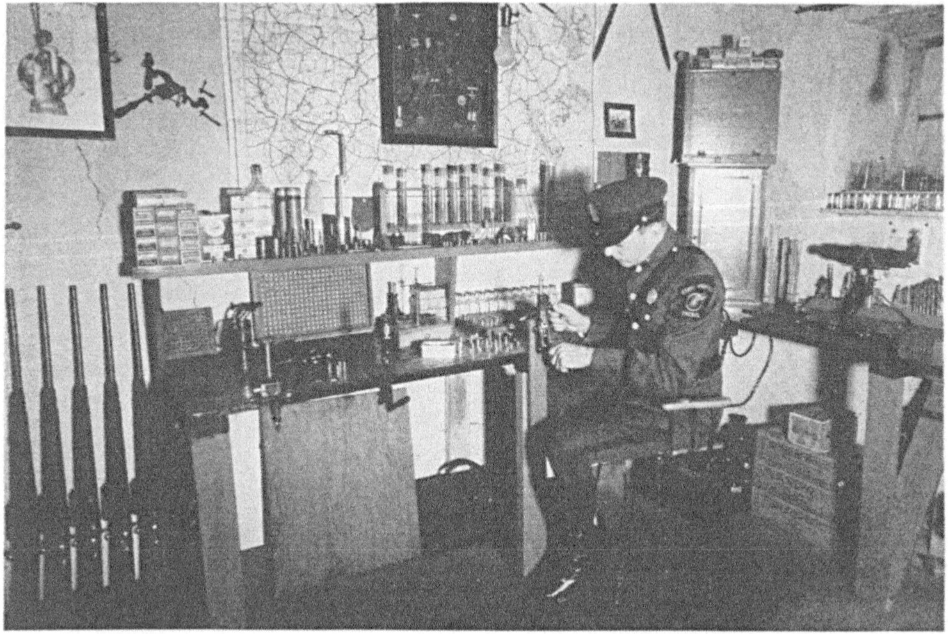

Pennsylvania Department of Highways

The author inspecting ammunition for the Pennsylvania Highway Patrol

MANUAL OF
POLICE REVOLVER INSTRUCTION

The information given in this MANUAL OF REVOLVER INSTRUCTION has been gathered from an extensive study of firearms, ammunition, experience as an Instructor of Small Arms in Police Departments, Military Organizations and Civilian Clubs.

The principles outlined are being used with success in Police Instruction and are definitely applied to practical Police requirements. The MANUAL is intended to bring the essential principles of revolver shooting to the Police Officer in a brief form.

BY

R. M. BAIR

Revolver Instructor and Ballistician
Pennsylvania Highway Patrol

NATIONAL RIFLE ASSOCIATION
BARR BLDG., WASHINGTON, D. C.

To

E. T. S.

CONTENTS

ILLUSTRATIONS

FOREWORD

This *Manual of Revolver Instruction* has been written to assist the *Police Officer* who realizes the necessity for better ***marksmanship*** and the ability to fire quickly and accurately should emergency arise, for experience has proven that his life or the lives of others may depend, at any moment, upon his skill as a marksman.

There are no *born shots*. However, with patience an officer may become a good revolver shot by following the methods of instruction herein given, providing he masters each chapter thoroughly and practices the methods as suggested in *dry shooting*. Every effort has been made to present the instruction by the shortest method. This system of ready reference to chapters makes this *Manual* a valuable assistant to the officer on the range or in the classroom.

Acknowledgment is made for valuable assistance rendered by Captain W. C. Price, Captain P. J. Dorr, Lieutenant E. T. Strange, Lieutenant C. M. Stroup, Pennsylvania Highway Patrol, Pennsylvania Department of Highways, Colt's Patent Fire Arms Manufacturing Company and others.

Hummelstown, Pa.
August 13, 1932.

Pennsylvania Department of Highways

Pennsylvania Highway Patrol Range, at Harrisburg, Pa.

INTRODUCTION

There are six definite *Principles* of revolver shooting that must be mastered before the officer will become a good revolver shot and get beyond the novice class. They are given in the order in which they are carried out when on the range in practice firing or firing whenever emergency demands. The officer's skill in shooting the revolver will depend largely upon how well he has mastered these *Principles* and to what extent he is able to concentrate on them at the time of firing.

The six *Principles* of revolver shooting are as follows:

POSITION
GRIP
SIGHTING
BREATHING
NO MUSCLE STRAIN
TRIGGER SQUEEZE

Each *Principle* will be carefully discussed under its separate chapter, with practical suggestions in practicing the *Principles* so that they will be easily mastered by the officer.

After continued practice, the officer will be able to carry out the *first five Principles* without much thought. In fact they are carried out by the sub-conscious mind in the same manner that he would release the clutch when driving an automobile. The sub-conscious mind plays a very important part in revolver shooting, providing it is properly trained, and will function quickly and correctly when needed.

POSITION

Before describing the proper *Position,* it will be necessary to establish two definite lines, so that the proper *Position* of the feet may be determined. They are namely the ***Firing Line*** and the ***Line of Fire.*** ***The Firing Line*** is the line upon which you stand at the time of firing. It is a measured distance from the face of the targets and runs parallel to the target line. The ***Line of Fire*** is an imaginary line from the muzzle of the revolver to the face of the target, and runs at right angles to the ***Firing Line.*** Fig. 1.

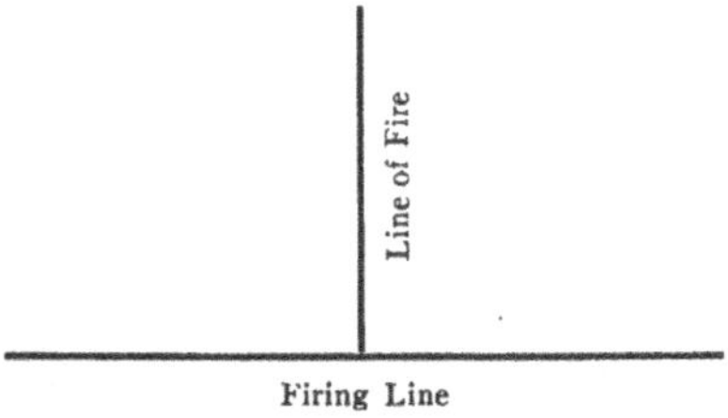

FIGURE 1

To assume the proper *Position* of the feet, if firing with the right hand, step forward with the right foot, pointing it 45 degrees to the left of the ***Line of Fire.*** The left foot is placed 18 to 20 inches to the rear, the heel several inches forward of the ***Line of Fire*** and the foot pointing parallel with the ***Firing Line.*** Fig. 2.

To assume the proper *Position* of the feet, if firing with the left hand, step forward with the left foot, pointing it 45 degrees to the right of the ***Line of Fire.*** The right foot is placed 18 to 20 inches to the rear, the heel several inches forward of the ***Line of Fire*** and the foot pointing parallel with the ***Firing Line.*** Fig. 3.

The officer should sketch the two lines (*Firing Line* and *Line of Fire*) on the floor and mark the location where the feet should be placed for the right and left hand position of firing. He should now assume the position of *Attention* standing on the *Firing Line* at the point the

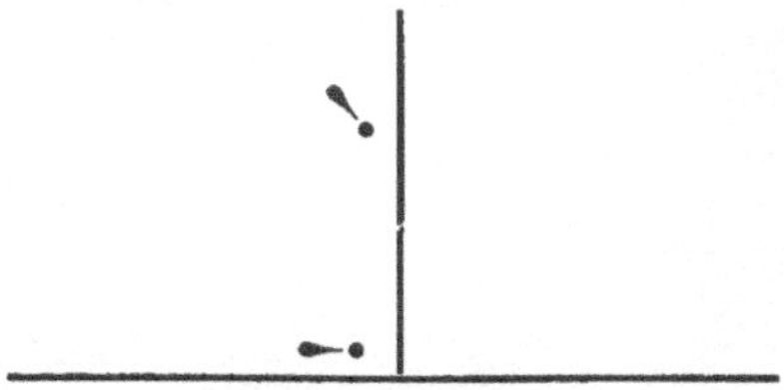

Right-hand position

FIGURE 2

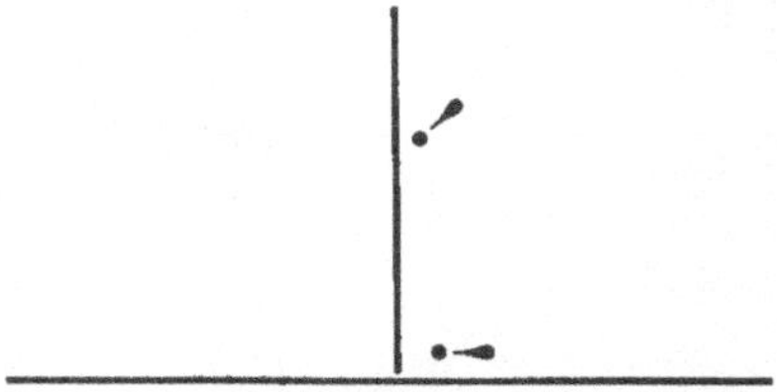

Left-hand position

FIGURE 3

Line of Fire intersects, facing the target. At the command *Fire* (which he can give himself) he should practice swinging into the proper position until he can assume the correct position, if firing with either hand, quickly and with ease. Should emergency arise or the command *Fire* be given on the range, by the Range Officer, he will then be able to carry out quickly and correctly the first *Principle* of revolver shooting. Fig. 4.

Several very good reasons for placing the feet in the proper position at the time of firing are as follows:

FIGURE 4.—Showing proper position of the feet when firing with the right hand. Left arm to hang at side of body. This position makes the body the smallest possible target and relieves *Muscle Strain.*

1. It will prevent *Body-sway,* and it will greatly assist *Holding* should there be a 9 or 3 o'clock wind blowing, for the reason that the officer has formed part of a triangle by placing his feet in this position and he will not sway very readily over the hips.

2. It makes the body the *Smallest Possible Target* to the adversary. This is a very important factor to the officer, for if he faced the adversary fully, he would present a larger target.

3. By assuming the proper position, the officer will prevent *Muscle Strain.* If he faced the target or adversary fully, he would cause strain in the back, shoulder and arm muscles, making him very unsteady in his *Hold.*

QUIZ

1. Name principles of revolver shooting in their proper order.
2. What is meant by (a) Firing Line? (b) Line of Fire?
3. For what purpose are the two lines (Firing Line and Line of Fire) established?
4. Describe the proper position of the feet when firing with the (a) Right hand. (b) Left hand.
5. What causes body-sway at the time of sighting and firing a revolver?
6. Give three reasons for standing in the proper position at the time of firing a revolver.
7. After the command *Fire,* what is the first principle to think of if firing with the left hand?

GRIP

The second principle in revolver shooting to be considered is *Grip*. The officer should master the proper *Grip* with both the right and left hand, for occasion may arise when it would be impossible for him to fire with the right hand owing to wound or injury to the hand

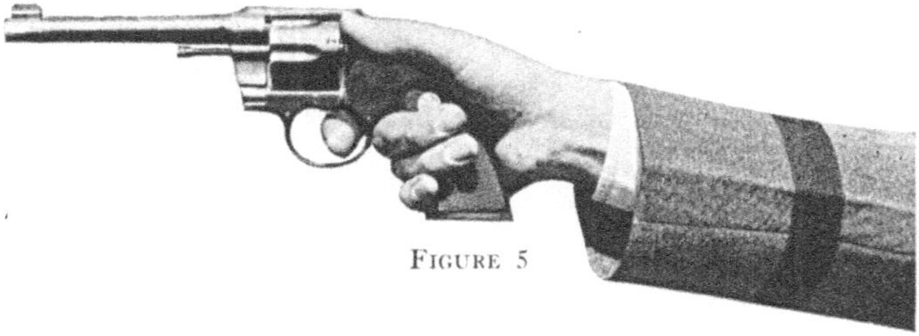

FIGURE 5

or arm. Without this ability to fire ambidextrously, he would only be half efficient, and would be placed at a disadvantage when firing from the corner of a building.

When firing with the right hand, grip the stocks with the last three fingers and the back of the palm, finger-tips not touching the stocks,

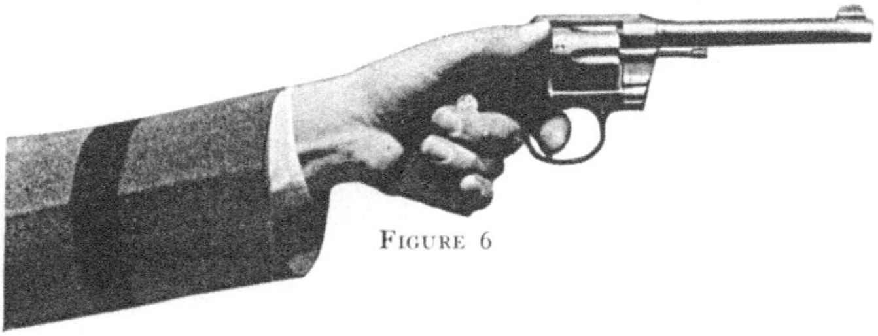

FIGURE 6

thumb placed on the latch and against the left side of the frame, beside the hammer, thumb-tip pointing upward and the first or trigger-finger placed on the trigger, wherever it touches. Fig. 5.

When firing with the left hand, grip the stocks with the last three fingers and the back of the palm, finger-tips not touching the stocks, thumb placed on the right side of the frame, following the contour of the top of the frame beside the hammer, thumb-tip pointing upward and the first or trigger-finger placed on the trigger wherever it touches. Fig. 6. Care should be taken that the trigger-finger is never placed high on the trigger, for the reason that the trigger pull will be increased. The trigger-finger should be placed on the rounded portion of the trigger provided for it.

The *Grip* should always be high on the stocks so the back of the hammer, when cocked, will touch the top part of the hand between the trigger-finger and thumb. This will cause the barrel of the revolver to point straight as naturally as pointing the finger, and the barrel will be in prolongation with the arm that is holding the revolver. It will also help to maintain the same *Grip* which is very important. If the officer were to change the *Grip* with each string of shots, it would change the *Group* on the target each time. A low *Grip* is hard to keep uniform for each shot and will cause wild shots on the target, for the reason that when recoil takes place, the muzzle of the revolver rises, causing the stocks to slip in the gripping hand after each shot is fired.

FIGURE 7

Another reason for gripping the stocks high, is to use the thumb-tip as a sight for *Snap* shooting or shooting in the dark, when it would be impossible to see the iron sights of the revolver. If the proper grip has been taken, it will bring the thumb-tip about level with the top of the rear sight, (the thumb-nail can be seen when it is dark, with the arm extended) and with the barrel pointing straight, it is possible to hit a man-sized target as far as the target is visible up to 50 yards or more. Fig. 7. This method of firing may be used with either hand and is very effective.

There are several *DON'TS* concerning *Grip* and care must be taken that they are avoided when firing, even though the hand has been placed properly on the stocks of the revolver.

1. Finger-tips should *never* touch the stocks.
2. Do not *Grip* too *tightly*.
3. The arm muscles should *never* be tensed or strained.

These mistakes in gripping will cause *Muzzle Wabble* resulting in wild shots. A large number of persons, when asked to fire a revolver, will say they are nervous. In most cases it is not nervousness but rather *Muzzle Wabble*.

For *Slow Fire*, the revolver should be gripped only tight enough to *Hold*, keeping the sights in alignment until the shot is fired by *SQUEEZING THE TRIGGER*. For *Timed* and *Rapid Fire* the grip is slightly tighter so that the same grip is maintained for the *string* of shots. The arm should be fully extended from the body at the time of sighting and firing but there should be no stiffness of the muscles.

The revolver should be fired *single action*, that is, to cock the hammer between each shot. The proper method of cocking the hammer with the right hand is to turn the muzzle to the right, cock the hammer with the thumb, close the gripping hand thus swinging the muzzle back to the left and bring the sights into alignment for the next shot, being sure to release the finger tips from the stocks before the shot is fired. To cock the hammer if firing with the left hand the method is the same except the muzzle is turned to the left when cocking with the thumb of the left hand.

Care must be taken at the time of cocking the hammer, that the wrist is not turned or rolled. Only the muzzle of the revolver is turned to the right or left as the case may be, but the sights of the revolver are kept in the vertical position when the revolver is turned.

Firing *double action* is to cock the hammer by increasing the pressure of the trigger-finger on the trigger, which lifts the hammer until it is released by the end of the *strut*. Fig. 8. Firing double action increases the trigger pull many times, for the reason that sufficient trigger pressure is required to depress the *main spring*, lift the *rebound lever* and *hand* that engages the *ratchet* on the *cylinder*, revolving the cylinder for the next shot and lifting the hammer. The revolver should only be fired by this method in emergency, when the hammer cannot be cocked by the thumb or when in very close proximity to the adversary, such as a hand to hand scuffle. At greater distances, firing double action will cause muzzle wabble, *flinching* and wild shots. It is impossible to keep the sights in proper alignment when the target is at greater distances.

When firing *Timed* or *Rapid Fire* by the *single action* method, care should be taken to release the trigger after each shot has been fired, otherwise, after the hammer has been cocked for the second shot and the trigger *SQUEEZED,* the hammer will again fall on the last cartridge fired. At the time the trigger has been released after a shot has been fired, the *rebound lever* forces the *trigger* forward

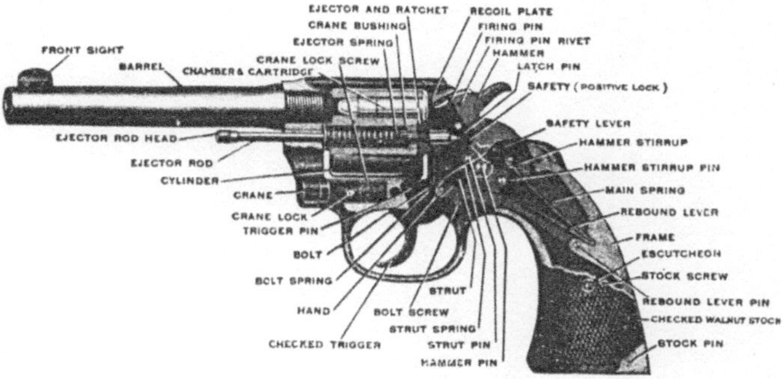

FIGURE 8

and allows the *hand* to drop so that it will engage the *ratchet* on the *cylinder* and when the hammer is again cocked it will turn the cylinder enough to bring the next cartridge primer into alignment with the firing-pin hole.

QUIZ

1. Describe the proper *Grip* when firing with the (a) Right hand. (b) Left hand.
2. Why should the finger-tips never touch the stocks at the time of sighting and firing a revolver?
3. Why should you never *Grip* too tightly at the time of sighting or firing a revolver?

Pennsylvania Department of Highways

The author instructing class of students in the proper method of loading. Training School, Pennsylvania Highway Patrol

4. How tightly should the stocks be gripped at the time of firing?
5. What is the advantage of placing the thumb high along the side of the frame when firing?
6. Should the *Grip* be high or low on the stocks at the time of firing?
7. How should the hammer be cocked, between shots, when firing with the (a) Right hand? (b) Left hand?
8. Why should the arm muscles never be trained or rigid at the time of sighting and firing a revolver?
9. Why should the trigger-finger never be placed high on the trigger at the time of firing?

SIGHTING

The third *Principle* to be thoroughly mastered by the officer is *Sighting,* which will require patience and honest effort in order that sighting problems may be worked out instantly at the time of firing.

FIGURE 9

FIGURE 10

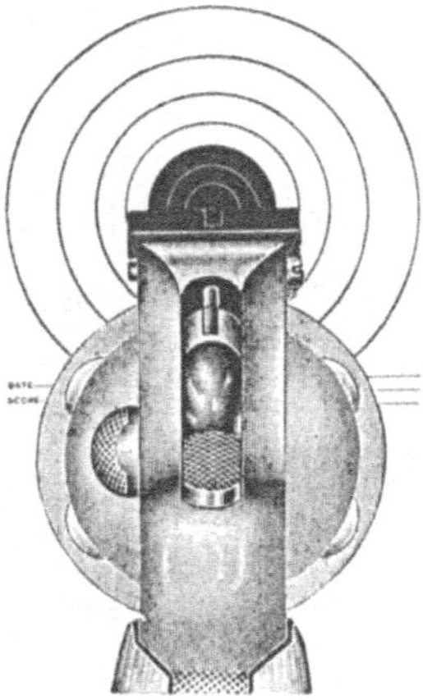

FIGURE 11

Without a thorough knowledge of sight alignment with relation to the target at the time the shot is fired, and the corresponding location of the shot on the target, the officer will never be able to account for wild shots.

The sights referred to in this manual are the *Patridge* and the *Blade* fixed types and are not adjustable, that is, they are part of the barrel and frame of the revolver and cannot be adjusted for *windage* or *elevation*. These types are used almost entirely by Police Departments, and have proven the most practical for holster or pocket use. The only difference between the two types is that the blade type front sight is thinner, and the rear sight notch is narrower. The patridge front sight and rear sight notch are both wider, making it easier to bring them into alignment quickly, and helps to overcome a natural tendency to *cant* the revolver when sighting. Figs. 9 and 10.

The reason for using the bottom edge of the *Bull's-eye,* or 6 o'clock, at the bottom edge of the 8 ring on the Standard American Target, is always to take the same *aiming point*. It also makes it possible to see the lines of light on either side of the front sight in the rear

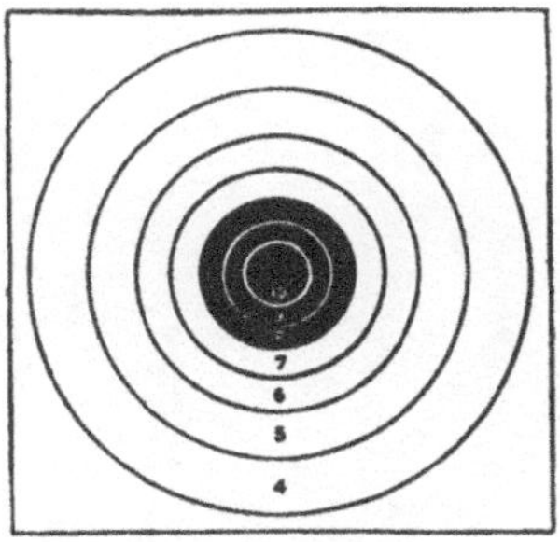

FIGURE 12

sight notch more clearly than if you used an *aiming point* inside the black portion of the target. The iron sights on the revolver are black and in certain light do not stand out distinctly if aimed at the black portion of the target, or what is known as a *Deep black hold*. However, some shooters prefer the *deep black hold* as shown in Fig. 11, especially if using a revolver equipped with the patridge type sights.

The Standard American Target is used almost entirely by Police Departments for training purposes or quarterly firing, for the reason that the inside diameter of the 5 ring is approximately the size of a man's body. It is used in most competitive matches, including the

National Matches. The diameters of the rings into which the target is divided are as follows:

10	ring	3.36	inches
9	"	5.54	"
8	"	8.	"
7	"	11.	"
6	"	14.8	"
5	"	19.68	"

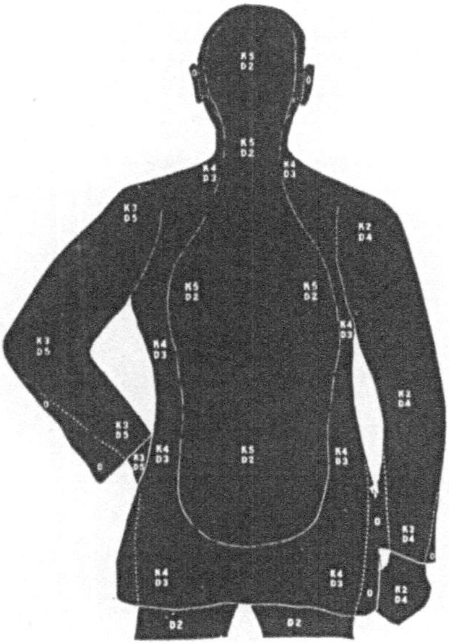

FIGURE 13

The rest of the target, 24 x 24 inches, count a value of 4. Fig. 12. Another very good target that is used by many Police Departments for rapid fire is the Colt's Police Silhouette Target. Fig. 13.

The location of a shot on the target is read by the *clock method* and the *numbered rings* on the target. That is, you read the face of the target the same as you read the time on the dial of a clock. All targets are composed of numbered rings that increase in diameter

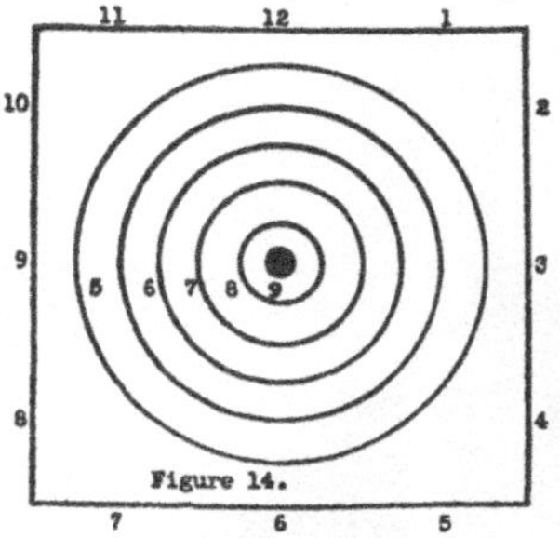

FIGURE 14

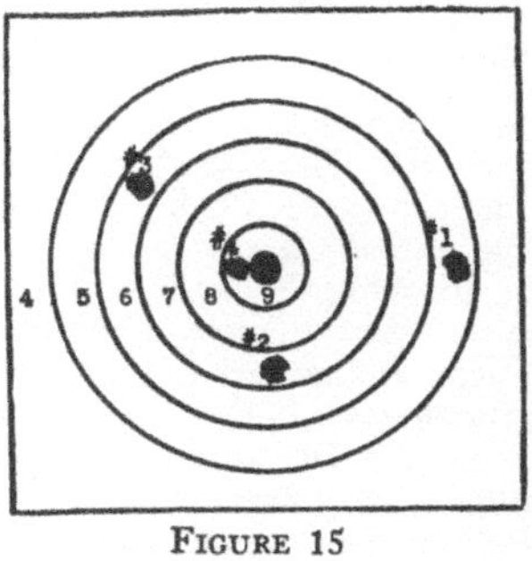

FIGURE 15

from the center of the target to the outside margin. Figs. 14 and 15.

Figure 15 shows the location of four shots on the target as follows:

Number one shot strikes the 5 ring at 3 o'clock.
Number two shot strikes the 7 ring at 6 o'clock.
Number three shot strikes the 6 ring at 10 o'clock.
Number four shot strikes the 9 ring at 9 o'clock.

The proper alignment of sights at the time of firing or at the time the *firing pin* strikes the *primer* of the cartridge and before the revolver goes into recoil, is to have the top of the *front sight* level with the top of the *rear sight,* and an equal amount of light on either side of the *front sight* in the *rear sight notch.* Figs. 16 and 17. The *bull's-eye* or black portion of the target should appear to be resting on the top of the *front sight,* which will make the *aiming point* 6

FIGURE 16

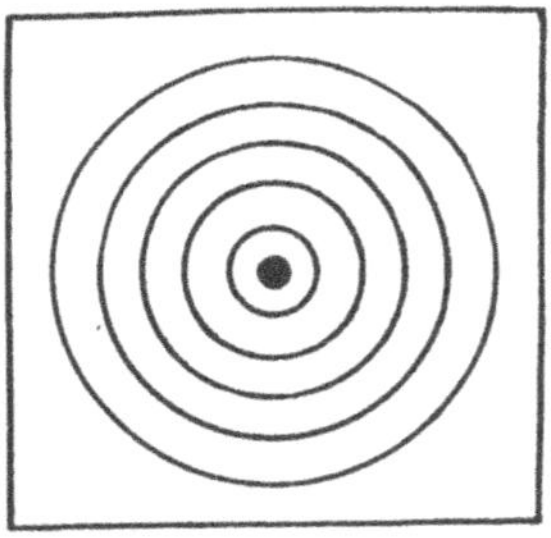

FIGURE 16a

o'clock at the bottom edge of the 8 ring or 6 o'clock black, on the Standard American Target. Figs. 12, 16, 16a and 17.

When sighting, no allowance should be made on the target for recoil when the shot is fired, as the factory has already taken care of this allowance by adjusting the height of the front sight in relation to the rear sight, to make provision for the slight upward movement of the barrel, from the time the ball starts from the *chamber* in the *cylinder* until it leaves the muzzle of the barrel. This adjustment of sights may be seen by inverting the revolver and resting the top of the front and rear sights on a level surface. It will be seen that the barrel is then pointing upward. If a revolver were clamped in a *fixed rest* with the sights in perfect alignment on a target at 25 yards, and fired, it would be found that the ball strikes 21½ inches below the center of the target or at the point the revolver was sighted, which shows the adjustment the factory has made in the alignment of sights to compensate for recoil at the time of firing. Fig. 18.

FIGURE 17

(Regarding adjustable sights; if the officer should happen to be using a target revolver with adjustable sights and wishes to move the sights to change the group on the target, a very good rule to remember is as follows:)

Whatever you do to the rear sight, you do to the bullet, that is if you were to move the rear sight to the right, the bullet will go to the right on the target.

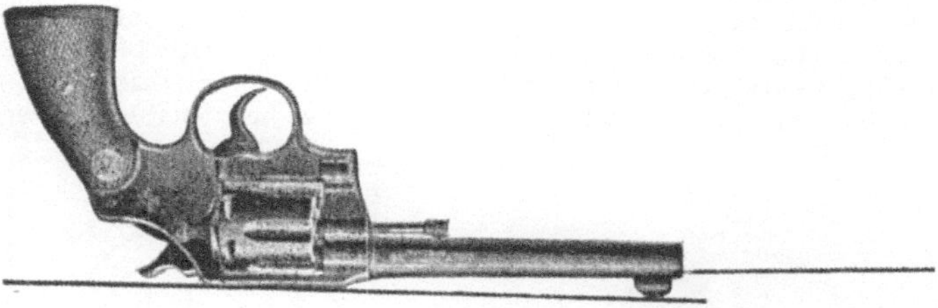

FIGURE 18

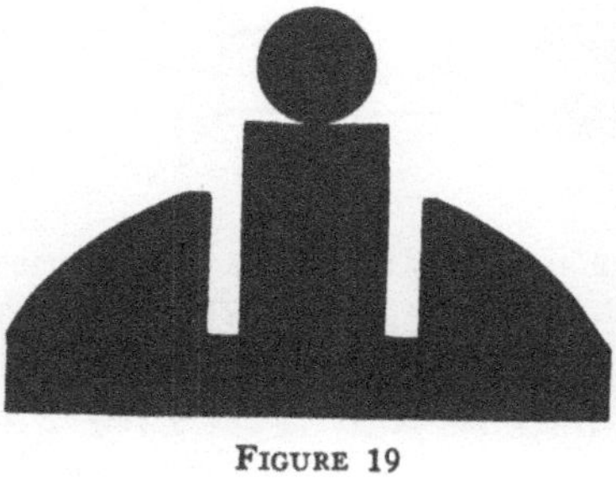

FIGURE 19

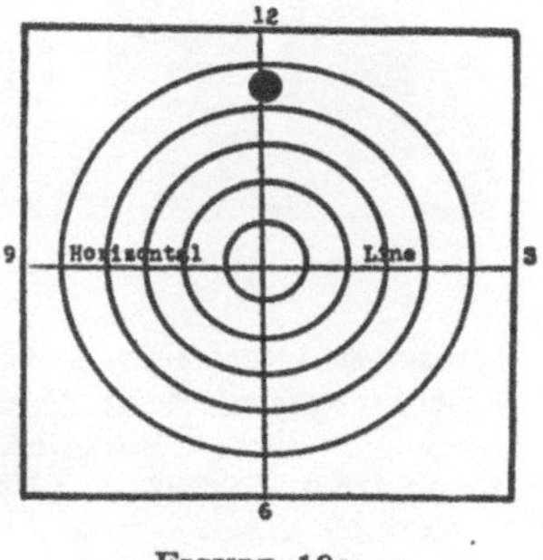

FIGURE 19a

Whatever you do to the front sight, will be opposite from where the bullet strikes the target, that is, if you were to lower the front sight, the bullet will strike high on the target.

At the time of sighting, the height of the front sight in relation to the height of the rear sight determines whether a shot will strike above or below the horizontal line drawn across the center of the target, from 9 o'clock to 3 o'clock. Figs. 19 and 19a. Any variation of light on either side of the front sight, in the rear sight notch, will determine whether a shot strikes to the right or left of the vertical line drawn up through the target from 6 o'clock to 12 o'clock. Figs. 20 and 20a.

Example No. 1.—If a shot strikes the target in the 5 ring at 12 o'clock, it shows that the top of the front sight was higher than the top of the rear sight, but that there was an equal amount of light on either side of the front sight, in the rear sight notch. Figs. 19 and 19a.

In this case, the top of the front sight, being higher than the top of the rear sight, caused the shot to strike above horizontal line drawn

FIGURE 20

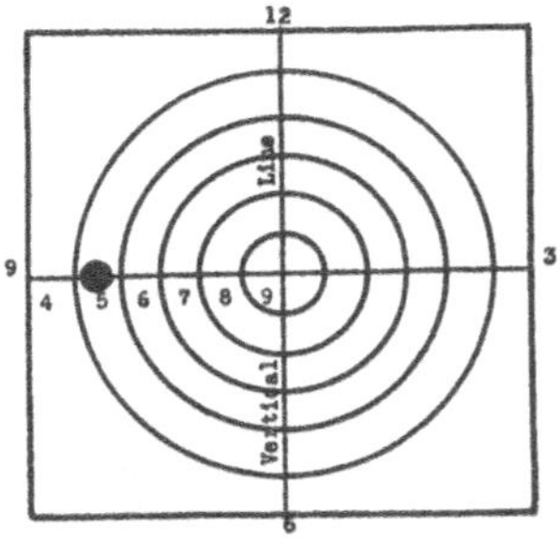

FIGURE 20a

across the target from 9 o'clock to 3 o'clock, but the lines of light on either side of the front sight, in the rear sight notch being equal, caused the shot to strike on the vertical line drawn up through the target from 6 o'clock to 12 o'clock.

Example No. 2.—If a shot strikes the target in the 5 ring at 9 o'clock, it shows that the top of the front sight was level with the top of the rear sight, but that there was too much light on the right side of the front sight, in the rear sight notch. Figs. 20 and 20a.

In this case, the top of the front sight was level with the top of the rear sight, causing the shot to strike on the horizontal line drawn across the target from 9 o'clock to 3 o'clock, but there was too much light on the right side of the front sight, in the rear sight notch, causing the shot to strike on the left of the vertical line drawn up through the target from 6 o'clock to 12 o'clock.

Example No. 3.—In figure 21a, we have a shot striking the 4 area at 1 o'clock. In this case the top of the front sight was higher than the top of the rear sight, and there was too much light on the left side of the front sight, in the rear sight notch.

Any shots striking to the right or left of the vertical line drawn up through the target, from 6 o'clock to 12 o'clock, will be known as a sighting problem in *WINDAGE*. Any shots striking the target above or below the horizontal line drawn across the target from 9 o'clock to 3 o'clock, will be known as a sighting problem in *ELEVATION*.

FIGURE 21

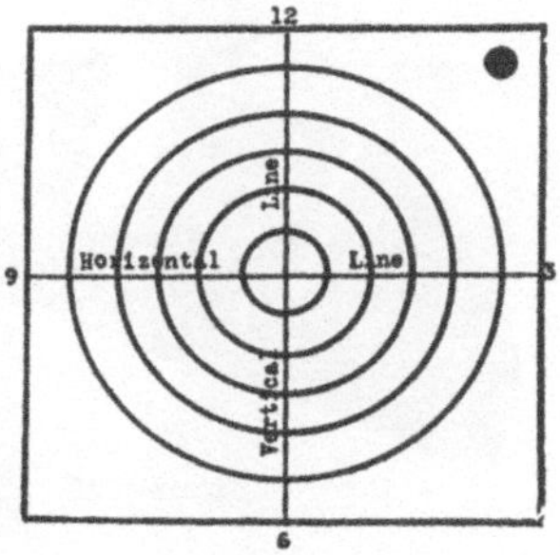

FIGURE 21a

In example number 3, we have two problems in sight adjustment, *windage* and *elevation*. The top of the front sight being higher than the top of the rear sight, caused the shot to strike above the horizontal line, drawn across the target from 9 o'clock to 3 o'clock. There was too much light on the left side of the front sight, in the rear sight notch, causing the shot to strike the target to the right of the vertical line, drawn up through the target from 6 o'clock to 12 o'clock.

Care should be taken to see that the revolver is never *CANTED* at the time of firing. This will cause the shots to strike high or low, right or left of the center of the target, depending upon the variation of sights in their alignment at the time of firing along with the degree to which the revolver is *canted.* Figs. 22, 22a, 23, 23a.

THE REVOLVER SHOULD ALWAYS BE HELD SO THAT THE SIGHTS ARE IN A VERTICAL POSITION AT THE TIME OF FIRING. Figs. 16, 16a and 17.

After the officer has mastered sight alignment and the corresponding location of the shot on the target, he should be able to *CALL* the location of the shot on the target as soon as the shot is fired, even though he is too far from the target to see where it strikes. The

FIGURE 22

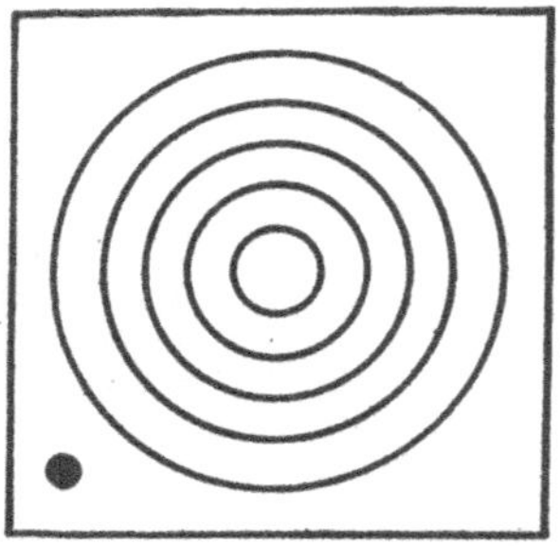

FIGURE 22a

ability to *call* his shots, as he fires them, is accomplished by seeing the alignment of sights at the time the hammer is released. There is a very short interval of time between the time the hammer is released and before the firing pin strikes the primer of the cartridge, so that during this interval of time, he should be able to get a picture of the alignment of his sights and their relation with the target. It is very important that the officer be able to *call* the location of his shots, for if he closes his eyes at the time of firing, his knowledge of sight alignment is useless to him, he will never be able to account for wild shots, and there will be a great tendency for him to *PULL THE TRIGGER.*

It is necessary that the officer learns quickly to sight the revolver with *both eyes open,* using the *master eye* to do the sighting; this is known as *BINOCULAR SHOOTING.* Nearly everyone has a *mas-*

ter eye, that is, one eye is usually stronger than the other. (In most cases it is the right eye.) To determine which is the *master eye,* point your finger at an object on the wall, keeping both eyes open. If you see two fingers and find that one appears more distinct than the other and is pointing directly at the object, the finger that appears more distinct is pointed by the *master eye.*

If the officer were pointing at a building or an object that happened to be lying on the floor, he would not close one eye, but would point with both eyes open. The same method should be applied when sighting and firing a revolver. Fig. 24.

FIGURE 23

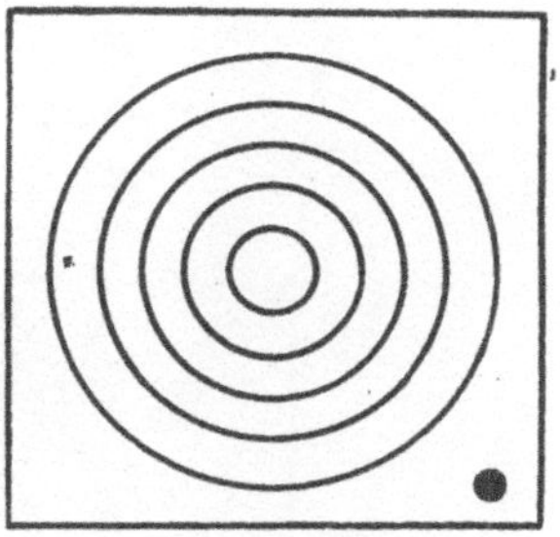

FIGURE 23a

It is of importance to the officer to keep both eyes open at the time of sighting and firing a revolver, in order to eliminate the following handicaps:

1. If one eye is closed at the time of sighting and firing, the officer will lose the vision to the left of him, if he were sighting with the right eye. This would be a serious matter to the officer, if he had to *cover* more than one person at one time, such as a number of persons in a room or during a riot.
2. Closing one eye will cause *squinting* with the other eye and in a very short time will develop eye strain.
3. Closing one eye will exaggerate any movement of the sights while sighting the revolver, which will have a tendency to cause *jerking,*

Pennsylvania Department of Highways

Pistol Instruction Class, Pennsylvania Highway Patrol

yanking and *pulling the trigger,* by trying to get the shot off the instant the sights appear in the proper alignment with the target.

4. By closing one eye, distances cannot be judged quickly and the target will not appear as distinct as when both eyes are open.
5. When one eye is closed, the sights will very often appear *fuzzy,* caused by eye strain.

Learning to shoot with both eyes open may be accomplished quickly by practicing the following simple methods:

FIGURE 24

Example No. 1.—Focus your eyes on a small object on the wall and then extend either arm fully with the thumb-tip pointing upward. Bring the tip of the thumbnail in line with the bottom of the object, by looking over the thumb-tip, and if you have been focusing your vision on the object, you will find that it is easy to see distinctly both the thumb-tip and the object and their relation, one with the other.

Example No. 2.—Now focus your eyes on the tip of the thumbnail and try to align it with the bottom edge of the same object. You will find that the object appears fuzzy, has a tendency to appear as though it were moving or two objects appear instead of one. For these reasons, always focus your eyes on the target or the object you intend to shoot at and then bring the sights into alignment. Always look *over* the sights and do not try to look at them or through them. Figs. 11 and 17.

Example No. 3.—Point the revolver at the target and bring the sights into alignment by closing one eye. (We will assume the left eye is closed and the sighting will be done with the right eye.) Now is the time to slowly *SQUEEZE* the trigger, and as you *SQUEEZE,* slowly open the left eye so that when the hammer falls, you have both eyes open. You can do this also by pointing the trigger finger at an object and slowly opening the left eye.

The eye that is the *master eye,* should do the sighting when firing with either hand. If you are firing with the right hand and the right eye happens to be the *master eye,* the revolver will be in prolongation with the right arm and eye. But if firing with the left hand and the right eye is the *master eye,* it will be necessary to have a slight bend in the wrist with the left arm extended to bring the sights into alignment with the eye and the target.

QUIZ

1. What is the proper alignment of sights at the time of firing a revolver?
2. Why should the revolver never be *canted* at the time of firing?
3. Why should you always *cock* the revolver, between shots, when firing slow, timed or rapid fire?
4. Describe the alignment of sights that would cause shots to strike the target as follows: (a) The 5 ring at 5 o'clock. (b) The 6 ring at 9 o'clock. (c) The 8 ring at 11 o'clock. (d) The 7 ring at 8 o'clock.
5. By what method do you read the location of a shot on the target?
6. If you were too far from a target to see where a shot strikes (a) Could you *call* its location? (b) By what method?
7. How would you sight a revolver at night to hit a man-sized target at fifty yards?
8. Should you sight with one eye closed or both eyes open?
9. Give three reasons for sighting with both eyes open.

10. Should the eyes be focused on the target or sights at the time of firing?
11. What is the difference between the *blade* and *patridge* type sights?
12. Why is the Standard American Target used for Police instruction?
13. Why should no allowance be made for recoil at the time of firing a revolver?
14. What rule should be remembered when adjusting movable sights on a revolver?
15. What is meant by *Binocular* shooting?
16. By what methods should you determine which is the *master eye?*

BREATHING

Breathing is the fourth *Principle* to be considered and mastered as the officer progresses with his revolver training. It is very necessary for him to breathe properly at the time of sighting and firing, or he will again face handicaps that will stand in his way of becoming a good revolver marksman.

If he continues normal breathing during the time of sighting, there will be a continual up and down motion of the arm that is supporting the revolver, as the breath is inhaled and exhaled. This motion will make it impossible for him to keep the sights in alignment with the *aiming point* on the target. When this happens, the novice shooter usually sees the sights in perfect alignment as they pass the *aiming point* on one of the up and down movements, and the trigger is given a *jerk* or *yank* in order to get the shot off. The resulting shot will be wild and very often a complete miss on the target, because of the *jerk* on the trigger.

If he takes a deep breath and holds it in the diaphragm, the result will be unsteadiness in holding and *muzzle wabble*. He will be uncomfortable because of muscle strain and again *jerk* or *yank* the trigger to get the shot off, so that he may breathe normally again.

The proper method of breathing is to hold the breath during the time of sighting and until the shot is fired. This is accomplished by breathing normally until the sights come into alignment with the target and you start to *squeeze* the trigger, then the breath is locked in the throat by forcing the tongue up against the roof of the mouth, holding it until the shot is fired. This method should be used for slow, timed and rapid fire. In timed and rapid fire, during the time required to cock the hammer, there will be a chance to inhale and exhale a short breath quickly, but again lock the breath in the throat until the shot is fired.

It will require practice to master proper breathing at the time of firing, until it becomes a mechanical operation with the officer. A tyro shooter has a great tendency to forget all about proper breathing at the time of firing.

QUIZ

1. Should the breath be held at the time of firing?
2. Should you continue breathing normally at the time of firing?

3. If you take a deep breath and hold it until the shot is fired, what will be the result?
4. If you continue breathing normally while sighting what will be the result?
5. When is the breath held?
6. How is the breath held?

MUSCLE STRAIN

Muscle strain presents the fifth *Principle* of revolver shooting, and is of importance to the officer should occasion arise which demands: *GETTING INTO ACTION QUICKLY* and a *COOL HEAD*.

Muscle strain means to the officer, strained muscles and nerves at a tension, for the tension of muscles and nerves are closely associated, whether he is firing a revolver or making an arrest in line of duty. The absence of *muscle strain* means that he will be *AT EASE* under all conditions. He will be better able to carry out his duties and will be qualified to meet emergency when it arises.

Muscle strain has a definite action on the officer's ability to shoot quickly and accurately should he be required to fire his revolver. At the time of firing, it makes him unsteady, uncomfortable, ill at ease and his mind will be in a state of panic, resulting in *muzzle wabble, pulling or yanking the trigger,* and wild shots that miss the target or adversary.

Some of the causes for *muscle strain* at the time of firing are as follows:

1. Straining the muscles of the arm that supports the revolver.
2. Gripping the stocks of the revolver too tightly.
3. The weight of the body not distributed evenly on both feet.
4. Taking an improper position, such as facing the target or adversary fully.

After taking the proper position, the officer should turn his head and look directly at the target, *draw* the revolver from the holster, raise the arm that supports the revolver (the arm being fully extended from the body but the arm and shoulder muscles relaxed) only high enough to bring the sights into alignment. Never lower the head to see the alignment of sights, for this will cause *muscle strain.*

When firing with either hand, the hand that is not gripping the revolver should never be placed on the hip. This will raise the shoulder higher than the shoulder of the arm that is supporting the revolver and cause *muscle strain.* The hand that is not gripping the revolver should be left hanging by the side of the body, Fig. 4. The thumb of the hand hooked in the cartridge belt or the hand placed in the pocket.

When firing the revolver, the officer should always be *at ease.*

QUIZ

1. When firing with the right hand, why should the left hand never be placed on the left hip?
2. What is the best method of overcoming *muscle strain?*
3. What will be the result when firing, if the stocks are gripped too tightly?
4. Why should the weight of the body be distributed on both feet at the time of firing?
5. Name three causes for *muscle strain* at the time of firing?

Pennsylvania Department of Highways

Students firing, Pennsylvania Highway Patrol

TRIGGER SQUEEZE

Of the six *Principles* of revolver shooting, *TRIGGER SQUEEZE* is the most important, and should be thoroughly mastered by the officer as quickly as possible through practice and concentration at the time of firing, for his proficiency as a good revolver shot will depend largely upon his ability to *squeeze the trigger* for each shot.

The proper method of *squeezing the trigger* at the time of firing, (being sure the proper *grip has been taken*) is to equalize the pressure between the trigger-finger and thumb until the shot is fired. The action of the thumb, at the time of *squeezing the trigger,* is forward and against the frame of the revolver if firing with either hand, equalizing the pressure of the trigger-finger before and after the hammer is released and until the firing-pin strikes the primer of the cartridge. Another method of equalizing the pressure between the trigger-finger and thumb, is to place the thumb against the side of the frame as described in the chapter on *grip* with sufficient pressure to equalize the pressure of the trigger-finger. In this case the thumb is not moved forward during the time of *squeezing the trigger.*

If the thumb pressure is *NOT* equalized with the pressure of the trigger-finger, at the time of *squeezing* and until the shot is fired, it will cause a slight movement of the barrel at the time the hammer is released before the firing-pin strikes the primer of the cartridge, causing the shots to group to the right or left of the bull's-eye, depending upon which hand is firing the revolver, even though the trigger was *squeezed* by the trigger-finger but without thumb pressure. This is a very common fault with many shooters who usually blame the revolver for being inaccurate.

The officer must be vary careful at the time of sighting and firing, not to give the trigger a *yank* causing *trigger pull* at the time he sees the sights in perfect alignment with the target. With the sights in alignment with the *aiming point* on the target, he should *hold* as close as possible and maintain a steady *trigger squeeze* until the shot is fired.

The officer will find that the sights weave back and forth, up and down, during the time he is trying to *hold* them on the target. He should not allow this to bother him, for as he continues his sighting practice this weaving will gradually diminish to a very small area, but under all conditions during the weaving he should *squeeze the trigger* until the shot is fired. It will be readily understood that when the sights are weaving across the target they are moving in parallel motion and if the shot is fired it will be confined to a very small area on the target, depending of course, on the extent of weaving. However, if he sees the sights in perfect alignment and pulls the trigger, he changes the angle of sights in their relation to each other and the target, causing wild shots and complete misses.

The officer should *never* know the exact instant the shot will be fired, for each time he knows when the shot will be fired, he will close both eyes and *pull the trigger,* causing wild shots and *flinching.*

The trigger-finger should be placed in the curved portion of the trigger, (Figs. 5 and 6) and the action of the trigger-finger at the time of *squeezing* is straight to the rear. He should be very careful not to place the trigger-finger high on the trigger (this is a common fault with novice shooters) as it greatly reduces the leverage on the trigger and increases the pounds trigger pull of the revolver, causing yanking, jerking and very often flinching.

The officer will never be able to fire accurately until he has learned to concentrate on *squeezing the trigger* for every shot. If he were to fire one thousand shots each day, one thousand times each day he would have to think and concentrate on *squeezing the trigger.* The reason for this will be found in the chapter on flinching.

Squeezing the trigger should be mastered in slow fire, before attempting to fire timed and rapid fire, for again the proficient timed and rapid fire shooter is the one who has mastered *trigger squeeze.* The method of *squeezing the trigger* is the same for slow, timed and rapid fire, except that in timed and rapid fire the action of the trigger-finger and thumb is speeded up when the trigger is *squeezed.* This action should be smooth and even and should never be jerky in order to get the shots off within the time limit. Again the officer should *hold* as close as possible and always *squeeze the trigger.* A very good rule for the officer to remember, in slow, timed or rapid fire, is to never hold longer than it requires to bring the sights into alignment with the *aiming point* on the target and *squeeze the trigger.*

A very good way for the officer to satisfy himself that he is squeezing the trigger properly, and he should try this frequently in his *DRY SHOOTING,* is to balance a coin (a dime is most suitable), on the barrel of the revolver about one inch forward of the frame. Cock the hammer first, place the dime on the barrel and see if he can *squeeze the trigger* until the hammer falls without having the coin drop off. This he should be able to do with the arm fully extended in the firing position, and after he becomes proficient with this practice he is well on the way to becoming a good revolver shot, providing he *squeezes the trigger* with the revolver loaded.

QUIZ

1. What is the most important *principle* to think of at the time of firing?
2. What is the action of the thumb when firing with either hand?
3. What is the proper method of *squeezing the trigger?*
4. What will be the result when firing if the thumb pressure is not equalized with the trigger-finger?
5. What will be the result when firing if the trigger is pulled or yanked?
6. By what method should the revolver be fired, if the arm that is supporting the revolver weaves back and forth, up and down, during the time of sighting?
7. By what method should the revolver be fired if the officer sees the sights in perfect alignment with the aiming point on the target?
8. Should the officer know the exact instant the revolver will be fired?
9. What rule should the officer remember when *squeezing the trigger* in timed and rapid fire?
10. When *dry shooting,* how may the officer determine if he is *squeezing the trigger?*
11. Why should the officer be careful to see that the trigger-finger is placed in the curved portion of the trigger at the time of firing?
12. Would firing a great number of shots each day overcome the necessity of having to think of trigger squeeze at the time of firing?

FLINCHING

FLINCHING is the most difficult obstacle or handicap for the officer to overcome while learning to become a good revolver shot. *Flinching* is an involuntary movement of the arm and shoulder muscles, causing tenseness at the time of firing and is very often accompanied by jerking the head and closing both eyes at the time of the *flash, explosion* and *recoil.*

Flinching is sometimes caused by carelessness or lack of concentration, but in most cases is caused by an unconscious *fear* of the *flash, explosion* and *recoil* at the time of firing the revolver. Very often, the shooter will unconsciously move his arm forward (tensing the muscles and nerves) to meet the recoil. The effect of *flinching* will cause wild shots on the target and very often a complete miss on a man-sized target at short range.

The officer can determine if he is *flinching* and to what extent, by loading several chambers in the cylinder with empty cartridge cases and several with loaded cartridges, giving the cylinder a spin and without looking to see the arrangement of the loaded cartridges in the cylinder close it. He should then start firing as though each chamber held a loaded cartridge and when the hammer falls on an *empty,* he will be able to see the degree to which he is *flinching.* This will show him how much the sights have been deranged or moved out of alignment with the target.

A very good method of overcoming *flinching,* is by properly *squeezing the trigger* and being able to *call* the location of the shot on the target, keeping both eyes open and seeing the alignment of sights at the time the hammer falls and before the revolver goes into recoil. In order to become a good revolver shot, the officer must learn to concentrate as quickly as possible on *hold, squeezing the trigger* and *calling the shot* at the time of firing.

As *flinching* is usually an involuntary action or the action of the sub-conscious mind telling the shooter to be afraid of the *flash, explosion* and *recoil,* he will have to guard against it with every precaution by concentrating on the proper methods of revolver shooting.

QUIZ

1. Define *flinching?*
2. What causes *flinching?*

3. What will be the effect of *flinching* at the time of firing?
4. What methods can be used to overcome *flinching?*
5. What is a good method for the officer to determine if he is *flinching* at the time of firing?
6. How will *flinching* affect the accuracy of shots fired?
7. Why is it so necessary for the officer to be able to *call* his shots at the time of firing?
8. What causes *flinching* at the time of firing besides fear of the flash, explosion and recoil?

Pennsylvania Department of Highways

Instruction in the proper method of unloading, students Pennsylvania Highway Patrol

SAFETY RULES

The officer should train himself to be extremely careful at *ALL* times in the handling of his revolver or any firearms. There are *no accidents with firearms.* All supposed accidents have been caused by carelessness in some way or another and regardless of how familiar he becomes with firearms, he should always be extremely careful and carry out all *SAFETY* measures to the last *NTH.* The following are a few *SAFETY RULES* that will be well to observe:

1. NEVER handle, point or look over the sights of any firearms handed to you without first opening the arm to be sure it is not loaded.
2. NEVER glance into your pistol or revolver hurriedly and decide it is not loaded. Look once to see that it is not loaded and then look again to be sure you have not made a mistake.
3. NEVER load or cock a pistol or revolver on any range unless you are at the *firing point* and facing the target.
4. NEVER turn around to talk to any person behind you when at the *firing point* without first lifting out the cylinder of a revolver or locking back the slide of an automatic pistol, and laying either one down on the shooting bench before turning around.
5. NEVER leave the firing point with a revolver without lifting out the cylinder of a revolver or locking back the slide of an automatic pistol.
6. NEVER in case of misfire, open the revolver for at least ten (10) seconds and keep the revolver pointed toward the target.
7. NEVER fire a shot until you know your shots are in alignment with the target, as close as you are able to *hold* them. You are only wasting your time and ammunition if you do not take pains with every shot.
8. NEVER talk to a shooter when he is at the firing point.
9. If a pistol or revolver is handed to you or you should pick it up, always look to see that it is not loaded and then look again to be sure you have not made a mistake.
10. A revolver should be loaded with the muzzle pointing toward the ground about a yard away from the feet. NEVER load a revolver if anyone is standing in front of you.

11. A revolver should only be loaded on the range after you have taken your position at the *firing point* and after the command *LOAD* has been given by the Range Officer.
12. Always be careful to see that there is no obstruction in the barrel of a revolver such as a cleaning patch, heavy grease or oil or any foreign matter. For at the time of firing there is a great possibility that the barrel will be bulged or bursted, causing injury.
13. A shot should never be fired while the officer is running. In the case of a running gun fight, the officer should stop and fire deliberately.
14. If running with a loaded revolver drawn from the holster, the officer should be careful to see that the hammer is not raised or cocked, that the trigger finger is not on the trigger and that the muzzle of the revolver is pointed toward the object to be shot.
15. In a running gun fight where the fugitive turns to fire at the officer, the officer should fall to the ground (where he makes a very small target), grip the revolver with both hands, rest both elbows on the ground and fire deliberately.

CARE OF THE REVOLVER

The officer should keep his revolver in perfect working order at all times, for he never knows at what instant his life or the lives of others may depend upon the functioning of the arm. The following are a few suggestions on the care of the revolver.

1. As soon after firing as possible, all working parts should be thoroughly cleaned. Give particular attention to the inside of barrel and the chambers of the cylinder, hammer, inside of frame at top of breech block, inside of frame over the barrel where the barrel extends into the frame, under the ratchet at the breech end of cylinder and the entire outside surface of the revolver.
2. The materials that should be used in cleaning a revolver are, a hi-grade solvent with oil base, wood cleaning rod and patches for cleaning the inside of the barrel and chambers. The solvent should be used for cleaning the outside surface and a small amount should be dropped into the working parts through the hammer slot. A light coating of the solvent should be left on the inside surface of the barrel and the chambers after cleaning.
3. When the *latch* is released on a revolver, the cylinder should never be thrown out or let fall out as this will eventually bend the crane, throwing the chambers in the cylinder out of alignment with the breech end of the barrel. When this happens, lead will be peeled from each bullet that is fired and in a very short time will cause inaccuracy.
4. *FULL METAL PATCHED* (or what is commonly known as a *steel jacket*) bullet should never be fired in a revolver. They will eventually damage the rifling in a revolver barrel and are very dangerous in congested districts, on account of ricochet.

Pennsylvania Department of Highways

The author conducting a functioning test of ammunition

DRY SHOOTING

Be Sure the Revolver is Unloaded

After having studied and practiced the principles of revolver shooting as suggested in the several chapters and before going on the range to fire his revolver, the officer should be ready to practice *DRY SHOOTING*. That is, to fire the *empty* revolver at a target either on the range or in a room that is well lighted, placing a reduced target on the wall and *snapping dry* to see how well he has mastered the six principles and his ability to carry them out in their proper order after the command *FIRE* (which he should give himself) from *position* to *squeezing the trigger,* and until each shot is fired.

It is a very good idea for two officers to study the *manual of revolver shooting* and start their revolver training together, so that one may *coach* the other during dry shooting and on the range. This method will develop their skill more rapidly, for the officer acting as coach will be able to see mistakes made by the officer that is shooting and make the proper corrections.

The officer should first practice *drawing* the revolver from the holster with the right hand, firing one *dry* shot at the target and then passing the revolver to the left hand and fire the second shot, being careful to assume the proper *position* in each case and to see that the proper *grip* has been taken. He should continue this practice with the *empty* revolver until it *feels* perfectly natural in either hand, with looking to see that he has taken the proper *grip*. Should emergency arise, he will then be able to *draw* quickly, knowing the barrel of the revolver is pointing straight (as naturally as he would point his finger) by the *feel* of the revolver in his hand. See Figs. 5, 6 and 7.

The officer should start his *drawing* practice by being slow and deliberate with each movement, from the time of *drawing* the revolver from the holster until the shot is fired. He will have to guard against being over anxious to attain speed while learning to *draw,* for this will be developed as he continues his practice in the methods suggested.

After *drawing* the revolver from the holster, the officer should never raise the revolver over his head before bringing the sights into alignment with the target. This is a lot of *WILD WEST HOKUM*, and if his adversary were a good pistol or revolver shot it would be entirely possible for him to fire two shots at the officer before he could bring his sights into alignment.

Always *draw* the revolver from the holster, take the proper *grip* by the *feel* of the revolver in the hand (in emergency firing using the thumb-tip as a sight) and bring the sights or thumb-tip into alignment with the target or adversary by the shortest possible route. This is accomplished by extending the arm fully, in a natural motion toward the object he intends to shoot, as naturally as pointing his finger at the same object. The trigger-finger should never be placed on the trigger while *drawing* the revolver from the holster, for if the revolver were loaded and happened to stick in the holster, it is entirely possible the officer might shoot himself.

At the command *FIRE* the officer should now be able to carry out the first five principles of revolver shooting in about the count of ONE and as a mechanical operation. Any thought in the matter should be by the sub-conscious mind, which allows him to give all his attention to the sixth and most important principle, *trigger squeeze*.

When dry shooting, the officer should never *hold* longer than the time required to *squeeze* the shot off. He should avoid the bad habit of *holding* too long and *freezing*, which is only another form of *flinching* or being afraid to *squeeze* the trigger. If he will concentrate on squeezing quickly and with steady pressure as soon as the sights come into alignment, it will help him greatly with timed and rapid-fire shooting. Dry shooting should be confined entirely to slow fire, until he has learned to co-ordinate the *trigger squeeze* with the alignment of sights on the target until the shot is fired, before attempting to fire timed and rapid fire.

The officer should always align the sights with both eyes open, being sure to see the alignment at the time the hammer falls, call the location of the shot on the target and in dry shooting continue to *hold* and sight several seconds after the hammer falls. This will help him to overcome being jumpy on the range. Care must be taken to see that the thumb is never placed over the *hammer slot* (this mistake is very often made by a novice shooter as he applies pressure with the thumb to the side of the frame) as it retards the fall of the hammer, causing a mis-fire. The thumb is placed *AGAINST THE FRAME BESIDE THE HAMMER* and not on the top of the frame over the hammer-slot.

The next step in dry shooting is to fire five shots with the right hand and alternate with five shots with the left hand. (Learn to count your shots as you fire them, this is important when firing timed and rapid fire.) Continue this alternate firing until you have fired fifty shots with each hand. At first the arm and back muscles will become tired and there will be a tendency to become unsteady, but keep up this practice each day and in a short time, you will be able to *hold* steady and with *ease* during each *string* as the muscles become developed.

In timed and rapid fire practice or shooting, care must be taken to avoid becoming *jumpy* and *fighting time,* after the command *FIRE* is given by the Range Officer. Sometimes the fact of a stop watch in the hands of the Range Officer will bring on a case of *buck fever* and nerves with some shooters, only because they have not trained themselves for this course of fire.

The first requisite of a good timed or rapid-fire shooter is regular *CADENCE* in his firing, that is, shots being fired at regular intervals during the time allotted for the *string,* instead of taking six seconds for one shot and two seconds for the next (which is a sure indication of fighting time and being jumpy) the good timed and rapid-fire shooter will fire a shot each four seconds in twenty-second or timed fire, and fire a shot each two seconds in ten-second or rapid fire.

The proper method of developing perfect *cadence* in timed and rapid fire is for the shooter to count for each shot. When he is on the firing line ready for a timed or rapid fire string, he will be standing at *raised pistol* with the hammer cocked. At the time the Range Officer gives the command *FIRE* he should start counting (but not too fast) as follows:

1. TIMED FIRE, 20 seconds for 5 shots.

At the command *FIRE* given by the Range Officer the shooter starts counting to himself: One thousand, two thousand, three thousand, four thousand, for each shot. This will give him plenty of time to cock the revolver, align the sights and *squeeze the trigger* without becoming jumpy or fighting time. Care will have to be taken so that the trigger is not pulled at the end of counting four thousand, but rather he should practice *trigger squeeze* until the shot will be fired approximately at this time. Practice with a stop watch will soon teach the officer how fast to count.

2. TIMED or RAPID FIRE, 15 seconds for 5 shots.

The procedure is the same as for twenty-second fire except the count is: One thousand, two thousand, three thousand, for each shot.

3. RAPID or QUICK FIRE, 10 seconds for 5 shots.

Count: One thousand, two thousand, for each shot.

Student class firing, Pennsylvania Highway Patrol

Showing that the empty cylinder should be placed under the hammer and that only five cartridges should be placed in the revolver

If the officer masters this method of counting in his dry shooting, he will have no difficulty on the range with timed and rapid fire, and it will teach him the proper *cadence* of firing. A *string* of shots is any number of shots fired in succession without re-loading, usually five (5) shots constitute a *string* and two *strings* constitute a *score.*

QUIZ

1. What is the importance of *dry shooting?*
2. What is meant by firing *single action?*
3. What is the best method of learning to fire timed and rapid *strings?*
4. What is meant by *cadence?*
5. What is meant by firing *double action?*
6. Under what conditions should the officer fire *double action?*
7. How many seconds are allowed for each shot in timed fire?
8. How many seconds are allowed for a *string* of 5 shots in rapid fire?
9. What is meant by *fighting time?*
10. How may the officer be sure he has taken the right *grip* without looking at the revolver in his hand?
11. How should the revolver *point* after the officer has taken the proper grip?
12. How should the officer start his *drawing* and *dry firing* practice?
13. After the revolver has been *drawn* from the holster, why should the officer never raise it higher than is required to bring the sights into alignment with the target?
14. How should the officer carry out the first five principles of revolver shooting, after he has had sufficient practice in *dry shooting?*
15. How may the officer overcome being *jumpy* during timed and rapid fire when on the range?
16. Why should the officer be very careful to never place the thumb over the hammer slot at the time of firing?
17. How can the officer overcome the tendency of being unsteady at the time of firing?
18. What method should be used to develop perfect cadence?
19. How should the officer count for a *string* of 20-second timed fire?
20. How many shots usually constitute a *string* of shots?

QUALIFICATION AND MATCH COURSES

The following are a few qualification courses that are being used by different Police Departments for quarterly or monthly firing, and are adaptable to any Police Department depending on the amount of ammunition allotted for the purpose.

—1—

STANDARD AMERICAN TARGET

10 shots	Slow Fire	15 Yards	Right hand
10 "	" "	15 "	Left "
10 "	" "	25 "	Right "
10 "	" "	25 "	Left "
10 "	" "	50 "	Right or left hand
10 "	Timed "	25 "	Right hand
10 "	Rapid "	15 "	Right "

Slow Fire .... 5 shots in 30 seconds
Timed Fire .... 5 " " 15 "
Rapid Fire .... 5 " " 10 "

70 rounds of ammunition per man.

Possible score: 700 x 700.

—2—

STANDARD AMERICAN TARGET

5 shots	Slow Fire	15 Yards	Right hand
5 "	" "	15 "	Left "
5 "	" "	25 "	Right "
5 "	" "	25 "	Left "
10 "	" "	50 "	Right or left hand
10 "	Timed "	25 "	" " " "
10 "	Rapid "	15 "	" " " "

Slow Fire .... One minute per shot
Timed Fire .... 5 shots in 15 seconds
Rapid Fire .... 5 " " 10 "

50 rounds of ammunition per man.

Possible score: 500 x 500.

—3—

STANDARD AMERICAN TARGET

5 shots	Slow Fire	15 Yards	Right hand
5 "	" "	25 "	" "
5 "	" "	50 "	" "
5 "	Timed "	25 "	" "
5 "	Rapid "	15 "	" "

Slow Fire........................One minute per shot
Timed Fire........................5 shots in 20 seconds
Rapid Fire........................5 " " 15 "

25 rounds of ammunition per man.
Possible score: 250 x 250.

—4—

STANDARD AMERICAN TARGET
(9 and 10 ring black)

10 shots	Slow Fire	25 Yards	Right hand
10 "	Timed Fire	25 "	" "
10 "	Rapid Fire	25 "	" "

Slow Fire........................One minute per shot
Timed Fire........................5 shots in 20 seconds
Quick Fire........................5 shots in 10 seconds

30 rounds of ammunition per man.
Possible score: 300 x 300 (National Match Course).

—5—

ARMY L TARGET

10 shots	Slow Fire	25 Yards	Right hand
10 shots	Timed Fire	25 "	" "
10 shots	Quick Fire	15 "	" "

Slow Fire........................One minute per shot
Timed Fire........................5 shots in 20 seconds
Quick Fire........................5 shots in 11 seconds

Smith & Wesson Hand Guns....................................McHenry & Roper
Hunting Pronghorn Antelope..................................Bert Popowski
The Woodchuck Hunter..Paul Estey
Small Bore Rifle Shooting......................................Edward Crossman
Modern Bow Hunting...Hiram Grogan
Salt Water Fishing Is Easy.......................................Jerry Sylvester
Fishing Flies and Fly Tying.....................................William Blades
This Business of Exploring.............................Roy Chapman Andrews
Handbook for Shooters and Reloaders......................P. O. Ackley
Greatest Fishing..Joe Brooks
Hunting Our Biggest Game....................................Clyde Ormond
How To Hunt Deer...Edward Freeman
Burning Powder...Douglas B. Wesson
Practical Rifle Shooting..Walter Winans
Handloading Ammunition.......................................J. R. Mattern
Shooting...J. Henry Fitzgerald
Rifles & Rifle Shooting...Charles Askins
Mauser Rifles and Pistols..W. H. B. Smith
The Book of the Pistol & Revolver...........................Hugh B. C. Pollard

Made in United States
Cleveland, OH
05 November 2026

47664003R10038